THE ELEMENTS

Lead

Susan Watt

BENCHMARK BOOKS

MARSHALL CAVENDISH
NEW YORK

Benchmark Books
Marshall Cavendish Corporation
99 White Plains Road
Tarrytown, New York 10591

Library of Congress Cataloging-in-Publication Data
Watt, Susan, 1958-
Lead / Susan Watt.
p. cm. -- (The elements)
Includes index.
Summary: Explores the history of the useful metal lead and explains its chemistry,
its uses, and its importance in our lives.
ISBN 0-7614-1273-5
1. Lead--Juvenile literature. [1. Lead.] I. Title. II. Elements (Benchmark Books)

QD181.P3 W37 2001
546'.688--dc21   2001035406

Printed in Hong Kong

Picture credits

Front cover: Charles E. Rotkin/Corbis.
Back cover: Andrea Pistolesi/Image Bank.
AKG London: 4.
Ann Ronan Picture Library: 11.
Art Archive: Beethoven House, Bonn/Dagli Orti 22; Palazzo Vecchio, Florence/Dagli Orti 7.
Corbis: Charles E. Rotkin 6, 15, 30.
Ecoscene: Andrew Brown 25; Ian Harwood 13.
Image Bank: Andrea Pistolesi *iii*, 10; Jeff Hunter 21.
Mary Evans Picture Library: Explorer 9.
Pictor International: 26.
Science Photo Library: Arnold Fisher *i*, 12; Biophoto Associates 23; David Taylor 20 (*above*); Detlev van Ravenswaay 17;
Richard Megna/Fundamental Photos 20 (*below*); U.S. Department of Energy 16.
Sylvia Cordaiy Photo Library: J. Howard 24.
Travel Ink: Barbara West 18.
TRIP: H. Rogers 27; M. Barlow 14.
Werner Forman Archive: Ashmolean Museum, Oxford 8.

Series created by Brown Partworks Ltd.
Designed by Sarah Williams

Contents

What is lead?

L ead is a soft, gray metal in Group IV of the periodic table. Although this metal is soft, you will find that it feels very heavy. This is because it has a high density. Lead's density is more than 11 times that of water and about 1.5 times that of other common metals, such as iron. It is so weighty that the name we give to this element is very similar to the English word *load* and the French word *lourde*, which both mean "heavy."

Inside the atom

Each of the individual atoms (tiny building blocks) in lead are relatively heavy. This is because lead has a high atomic number. An element's atomic number shows how many protons (positively charged particles) the element contains in the nucleus at the center of the atom. Lead has atomic number 82—just 12 protons fewer than that of plutonium,

Lead has had all kinds of uses you might not imagine. In the 16th century, women such as Queen Elizabeth I used white cosmetics that contained lead.

ISOTOPE FACTS

This table shows the lead isotopes that occur in nature and gives the number of neutrons in their nucleus, their atomic mass, and the percentage of the isotope in an average lead sample.

Isotope	No. of neutrons	Atomic mass	% in average lead sample
^{204}Pb	122	204	1.4
^{206}Pb	124	206	24.1
^{207}Pb	125	207	22.1
^{208}Pb	126	208	52.4

which is the heaviest of all naturally occuring atoms.

Although all lead atoms have 82 protons, they can have different numbers of neutrons (uncharged particles) in their nucleus. This means that any piece of lead includes several different types of atoms, called isotopes. Each isotope of lead has a different atomic mass, which is arrived at by adding together the number of protons and the number of neutrons. On average, lead atoms have an atomic mass of 207.2, because of the different lead isotopes.

LEAD ATOM

Nucleus

First shell
Second shell
Third shell
Fourth shell
Fifth shell
Sixth shell

The number of positively charged protons in the nucleus of an atom is balanced by the number of negatively charged particles, called electrons, outside the nucleus. A lead atom contains 82 electrons. These orbit the nucleus in six layers, or shells. There are 2 electrons in the inner shell, 8 in the second shell, 18 in the third shell, 32 in the fourth shell, 18 in the fifth shell, and 4 in the outer shell.

DID YOU KNOW?

PLUMBUM

The chemical symbol for lead is Pb. This comes from the word *plumbum*, which means "lead" in Latin. The Romans (who spoke Latin) used a large amount of lead, particularly in their water systems. This is why we also use the word *plumber* for someone who installs water pipes.

Special characteristics

Lead is a naturally bright, shiny metal like silver or steel. You are unlikely to see it in this state, however, because, as soon as it is exposed to the oxygen in air, it develops a dull, bluish-gray covering. Unlike steel, lead is a soft metal—you could scratch it with your fingernail. This means that it is easy to form a piece of lead into different shapes. Lead also melts at the relatively low temperature of 622°F (328°C), and molten lead can be poured into a mold and left to harden in the shape of the mold. Lead is also resistant to corrosion (being attacked by chemicals) and can conduct electricity. Qualities such as these have made lead very useful, since the time it was discovered several thousand years ago.

Lead can be shaped very easily, because it is so soft and has such a low melting point. This man is pouring molten lead into molds to form lead ingots.

DID YOU KNOW?

LEAD PENCILS

The Romans discovered that it is possible to draw dark lines on paper using a piece of lead. For this reason, the center of a pencil is referred to as its lead—even though it has always been made from graphite (a form of carbon), and not lead at all.

The basest element

Before modern chemistry, people known as alchemists carried out experiments, in an attempt to turn different metals into one another and to find a substance that could prolong life. One of their aims was to make pure gold from other metals. The alchemists regarded lead as the oldest of all metals and also the most base (the farthest away from gold). The practice of alchemy also had a spiritual side—while gold was identified with goodness, lead was thought to represent sin and evil.

Today we know that it is not possible to make gold from other metals, and lead is regarded as a useful substance in its own right. Although lead cannot be made into gold, it can be used in many alloys

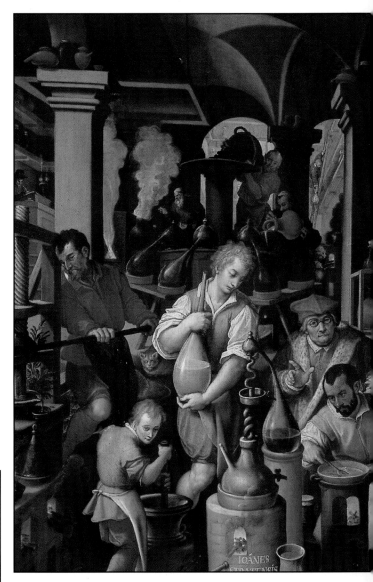

This painting by Flemish artist Jan van der Straet (1523–1605) shows an alchemist's laboratory in 1570. These people might be hard at work trying to turn base metals, such as lead, into gold.

(mixtures) with other metals. It is a substance that needs to be handled with care, however, because lead can be hazardous to health if it or its compounds build up in the body.

DID YOU KNOW?

CONDUCTING ELECTRICITY

Compared to other metals, lead is not a very good conductor of electricity. Most common metals conduct electricity very well. This is because their outer electrons are only loosely attached to the nucleus of the atom. Some of them break free, forming a "sea" of electrons. These electrons can then move through the metal, producing a flow of electricity. Among the elements, the metals copper and silver are the best electrical conductors. They conduct electricity about 12 times better than lead does. This is why you often see copper wires but you never see wires made of lead.

Lead in the ancient world

The ancient Egyptians used lead in pottery glazes thousands of years ago. This glazed statue of a lion was made in Egypt around 2686 B.C.

Lead is one of the few elements that has been known since ancient times. Archaeologists have found that lead was used by the ancient Egyptians at least 6,000 years ago—and possibly even several thousand years before that. In ancient Egypt, lead was used in pottery glazes and in making small statues and ornaments.

The Minoans were using lead by approximately 2000 B.C., at about the same time that lead coins came into use in China. In the Hanging Gardens of Babylon—one of the Seven Wonders of the Ancient World and probably created in the 9th century B.C.—lead sheeting was used on the floors, to hold the soil and keep in its moisture. Lead is also mentioned several times in the Old Testament of the Bible.

The Greeks and Romans

The ancient Greeks produced lead as a result of mining and refining silver, since both metals are found in the same rocks. They used sheets of lead as a protective covering for ships' hulls and for roofs—a practice that continues today.

By the time of the Romans, lead was being used in very large quantities.

The advanced plumbing and water supply systems built by the Romans contained lead piping. To make the pipes, sheets of lead were cut into rectangles and then bent around a wooden pole. The edges were joined with melted metal called solder, which also contained lead. During this period, lead continued to be used on roofs and ships, as well as in ornaments, drinking cups, and other everyday objects.

DID YOU KNOW?

COLORFUL LEAD

In the past, lead was often used in cosmetics. For example, ancient Egyptian women used a lead compound as a blusher, while women in 16th-century Europe used another lead compound to give them an attractive pale complexion. Naturally occurring lead compounds have often been used for their color, particularly white lead (lead carbonate) and red lead (minium, a lead oxide). These compounds were still being used in paints until a few decades ago.

Sweet lead

The Romans even used a lead compound in their cooking! They were fond of a sweetener called *sapa*, which they made by boiling sour wine in lead cooking pots. The vinegar (acetic acid) in the wine reacted with the lead in the pots to form white crystals of lead acetate, or "sugar of lead," which could be used to sweeten food. As we shall see later, lead is poisonous. The Romans used so much lead that some experts think lead poisoning may have been one of the main reasons why the Roman Empire came to an end.

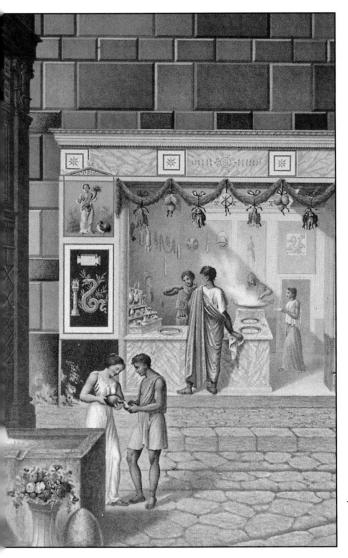

An ancient Roman scene in Pompeii, Italy, showing people collecting water from a fountain and buying food from a shop. Although the Romans realized that lead affected their health badly, they continued to consume it for centuries.

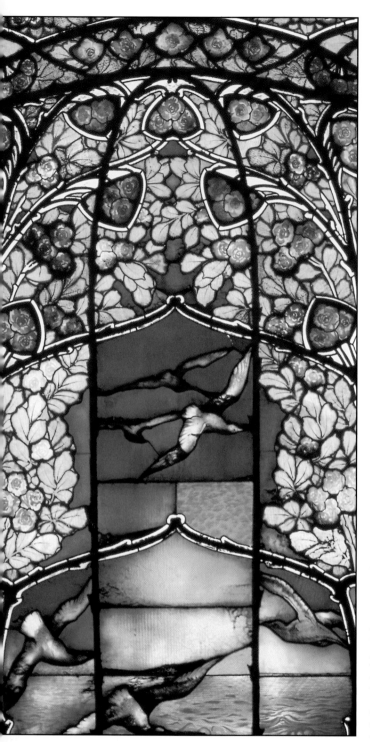

Stained glass is still used in some parts of the world. It is made by joining together small pieces of colored glass, using strips of lead. This creates a beautiful sparkling mosaic through which the light shines.

In the Middle Ages and beyond

Like the Romans, people in the Middle Ages found lead to be a useful substance, because it is easily rolled out and shaped or cast in molds, just as plastics are today. Lead was often used on the roofs of churches and cathedrals, to protect the buildings and to ensure that they would last as long as possible. The great Gothic cathedral at Chartres in France, built in the 13th century, had a lead-covered roof that survived more than six centuries in good condition. As early as the 7th century, thatched roofs in England and France were being replaced by lead, to guard against destruction by fire. Lead was also used in the intricate stained glass that began to be used in churches and cathedrals at this time.

Household pewter

Throughout the Middle Ages, materials containing lead were also used in homes. Pewter, a blend of tin with other metals including lead, had been used by the Romans for all sorts of household objects. After the Romans, pewter was used much less, except for church ornaments and pilgrims' badges. Around 1300, people began making pewter in large quantities again, and, by William Shakespeare's time, it was in common use in European households as drinking mugs, plates, cutlery—even chamberpots. Pewter is still made today, but now it contains no lead, and it is safe to eat and drink from it.

In medieval times, lead continued to be used in pottery glazes, as it had been across the ancient world. Medieval potters experimented with adding small amounts of other metals to the lead glazes, producing a variety of bright blues, yellows, and greens.

Printing

With the invention of printing in Germany in the middle of the 15th century, lead found a new use in the metal used to make the letters (called type) that printers used in their presses. Like pewter, the metal used was a blend of tin and lead. The exact blend was chosen so that it could be easily melted and cast in the intricate molds used for the letters but was hard enough for the

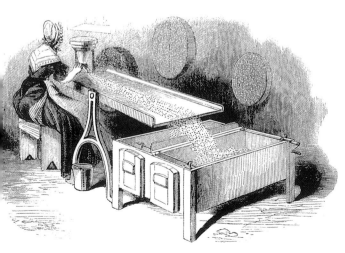

press. Over the centuries, the preferred composition for type-metal changed, but lead was always a main component.

Firearms

When people began to use firearms, they made simple bullets from lead. These bullets were called shot. The high density of lead in the shot ensured a powerful impact. An ingenious process was invented around 1650 for making lead shot. In this process, the molten metal was poured through a sieve into a pool of water far below. The metal droplets formed perfectly round balls as they fell and passed through the holes in the sieve, freezing solid when they hit the water.

This woman is sorting lead shot so that only the perfectly round bullets will be used. As the shot rolls down, the misshapen pieces move more slowly than the round pieces, so they fall into the box on the left, and the round pieces fall into the box on the right.

Where lead is found

Lead is found in smaller quantities on Earth than are metals such as iron and aluminum, but it is far more abundant than precious metals such as gold and platinum. On average, lead occurs in Earth's crust at a proportion of 12 parts in a million, making it about the 30th most abundant element on Earth.

Lead ores

Like most metals other than gold, lead itself does not usually occur naturally. Instead, it is found in compounds, combined with other elements, such as sulfur, oxygen, and carbon. These compounds do not occur in a pure form either, but tend to be combined with many other compounds in different proportions.

DID YOU KNOW?

LEAD AND SILVER

The ancient Greeks extracted lead and silver from mines at Laureion in eastern Greece. Although they obtained a huge amount of silver (which greatly added to the riches of the city of Athens), their methods were very inefficient by modern standards. Up to a third of the silver remained in the material left over after processing. For this reason, piles of reject material were reprocessed many times over the centuries that followed, and more and more of the precious metal was extracted.

Rocks that contain useful amounts of metals and other minerals are called ores. The most important lead ore is a rock called galena, which contains a large amount of lead sulfide (PbS). Galena usually also contains smaller amounts of other metal compounds, including those of zinc and silver. In fact, from ancient times to modern, the silver content of a lead ore has often been the most valuable part of it, despite the much greater proportion of lead. Other ores that contain lead include anglesite (lead sulfate, $PbSO_4$) and cerussite (lead carbonate, $PbCO_3$).

This mineral is anglesite, a naturally occurring form of lead sulfate. Anglesite is created when the lead sulfide deposits in galena react with oxygen.

Mining and recycling

More than 3 million tons (3.3 million tonnes) of lead are mined every year. China produces the most mined lead, followed by Australia, and then the U.S. As much lead is obtained by recycling lead that has already been used as is produced by mining, so total lead production is around 6½ million tons (6 million tonnes) a year. Overall, the U.S. produces most of the world's lead—but it uses even more of this element than it produces.

Recycled lead has become more and more important in the last few decades. Most recycled lead comes from lead-acid batteries, which are used mainly in automobiles, but old lead pipes and sheets are also recycled. Using scrap in this way has made lead the most recycled material after the precious metals.

Today household recycling centers are common in the Western world. In old lead-acid batteries from automobiles (shown below), up to 95 percent of the lead can be recovered.

Extracting lead

The starting point for extracting lead is mining the ore from Earth's surface. Drilling and blasting machinery are used to perform this task. Once the ore has been mined and taken away to the processing factory, it is ground up so that the different minerals and rocks can be separated out. At the end of this processing stage, the lead has been greatly concentrated and makes up about 70 percent of the total material.

Smelting

The next stage of the extraction process is smelting, in which the lead compound in the concentrated ore is changed chemically into metal. If the starting ore is galena, this means the lead sulfide (PbS) must be changed to lead (Pb). This is done in two stages.

First the material is burned in a blast of air, which converts the lead sulfide to lead oxide and sulfur dioxide (which is a gas and can be blown away). Then the lead oxide is converted to lead metal by heating with carbon (in the form of charcoal) in a blast furnace. This is usually a vertical structure—the concentrated ore is put in at the top and the molten metal formed in the furnace flows out at the bottom.

When it has been mined from the rock, lead ore contains about 10 percent lead on average. It needs to be refined further to produce useful amounts of lead.

Refining

The lead now exists as a metal, but it is still impure because it contains tiny traces of many other metals, including copper, tin, silver, and gold. The next processing stages, called refining, remove the other common metals but still leave the precious metals in with the lead. The gold and silver are removed by adding zinc to the molten lead. The precious metals combine with the zinc and float to the surface, where they can then be removed and refined.

The process of lead extraction is still not completely finished at this stage. As lead needs to have different purities, depending on its eventual use, it is produced in the U.S. and around the world in four different grades of purity. The top grade must be at least 99.99 percent pure lead, while the lowest grade is 99.85 percent pure.

This man is overseeing the process of lead refining. Before the final impurities have been removed to form pure metal, the lead is known as lead bullion.

ATOMS AT WORK

Most lead produced from mining starts as lead sulfide. It is converted to lead by a chemical process called smelting, which takes place in two stages. When lead sulfide is heated in air, oxygen combines with it to form lead oxide and sulfur dioxide.

Oxygen

Oxygen
3x O_2

Lead

Lead oxide
2x PbO

Sulfur

Lead sulfide
2x PbS

Sulfur dioxide
2x SO_2

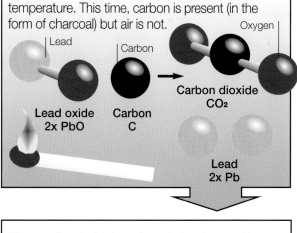

The lead oxide is then converted to lead metal by a second process, also carried out at high temperature. This time, carbon is present (in the form of charcoal) but air is not.

Lead

Carbon

Oxygen

Carbon dioxide
CO_2

Lead oxide
2x PbO

Carbon
C

Lead
2x Pb

The reaction that takes place during the smelting process can be written like this:

$$PbS + C + 2O_2 \rightarrow Pb + SO_2 + CO_2$$

The second stage, where oxygen is removed from lead oxide, is a reduction reaction. In this reaction, carbon is said to be the reducing agent.

Lead and radioactivity

The simplest substances that occur on Earth are called elements. Elements combine with each other to make compounds, but they cannot be changed into one another chemically. As lead is an element, the amount of it on Earth should stay the same. In fact, lead is continuously being produced, although only in relatively small quantities. This is because of the way that natural radioactive elements, such as uranium, change over time.

Radioactive elements are so called because they emit radiation. Their nuclei are unstable and shoot out particles or rays, which are detected as radioactivity. As they do so, the atoms change, or decay, into other elements.

Radioactive uranium

Uranium (element number 92) is the heaviest element that occurs naturally in significant amounts. When one of the uranium isotopes decays, it loses two neutrons and two protons—together these are called an alpha particle. As the uranium isotope decays, it becomes an atom of thorium (element number 90). The thorium atom is also unstable and decays to produce another unstable element, which also decays, and so on in a kind of chain. At the end of the chain, the stable element lead is produced. The succession of different radioactive elements and isotopes formed in this way is called a decay series.

There are several isotopes of uranium, all of which are radioactive. This picture shows a uranium isotope called ^{235}U, which is used as an explosive in nuclear weapons and as a fuel in nuclear reactors.

Decay series

There are three natural decay series, two of which begin with uranium. The third begins with thorium. Each series ends with a different isotope of lead. Scientists know all the steps in the decay series and how long each one takes. This means that lead can be used in dating rocks on Earth—or even those that

By dating the lead isotopes that naturally occur in meteorites, scientists can estimate the age of our Solar System. This meteorite was found in the Atacama Desert in Chile, in 1822.

land on Earth from space. Of the four different isotopes of lead that occur naturally, three are produced by radioactive decay.

The proportion of the different lead isotopes in a sample of rock depends on how long the decay has been going on—that is, the time since the rock was formed. So, by finding out the ratio between the different isotopes in a specimen of rock and knowing the rate at which the isotopes are formed, the age of the rock can be calculated.

How lead reacts

Chemically, lead is a typical metal. It is stable on its own but can react with other elements to form a wide range of compounds. However, one of the main reasons that people have found lead to be such a useful material for so long, is that it appears to be remarkably unreactive and does not corrode (wear away) with moisture.

Protective coating

Pure lead reacts very quickly when exposed to air, forming a thin coating of lead oxides. These substances do not dissolve in water, however, so they remain on the surface of the lead to protect the metal from further reactions. This is why lead has been so useful as a protective coating for roofs and the hulls of ships. Lead even resists attack by concentrated sulfuric acid and is used to line the containers in which this corrosive substance is stored.

Lead is so unreactive that seawater cannot damage it. It is used to coat underwater cables and ships' hulls (the main body of the ship, which sits in the water, painted red in this picture).

Reactions with acids

Lead does react with other acids, however. For example, it reacts slowly with hydochloric acid to form lead chloride ($PbCl_2$) and rapidly with nitric acid to form lead nitrate, $Pb(NO_3)_2$. It also reacts with organic acids (those that contain linked carbon atoms). This means that lead should not be used in the preparation of wine or fruit, as these foods contain organic acids (such as citric acid) that would react with the lead to form poisonous compounds. Despite this problem, as recently as

20 years ago, lead was still used to cover corks to form an air-tight seal on some wine bottles. Corks are porous (contain tiny holes), so the acids in the wine could react slowly with the lead, building up a high concentration of lead compounds if the wine was kept for a few years. For this reason, different materials are now used to seal wine bottles—or if lead is used, it is coated on both sides with a layer of tin to prevent contact with the wine.

Changing valencies

Like other metals, lead can react with nonmetals to produce a range of compounds, such as oxides, iodides, chlorides, sulfides, and nitrates. Lead has a changeable chemical nature, however. In its compounds, it often reacts as if it is able to form two chemical bonds, making the so-called lead II compounds. At other times, it reacts as if it can form four bonds, making the lead IV compounds. The number of chemical bonds an element can make is described as its *valency*. So lead has valencies of two and four. Lead IV compounds are generally less stable than lead II compounds.

Lead shares Group IV of the periodic table with another metal (tin), two semimetals (germanium and silicon), and a nonmetal (carbon). Not suprisingly, then, it can also form some compounds that are more typical of nonmetals. Examples of

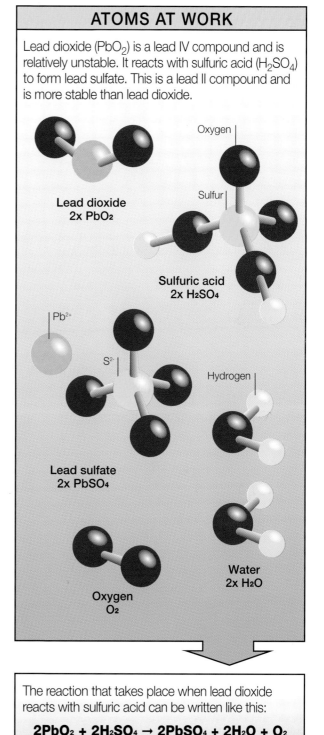

ATOMS AT WORK

Lead dioxide (PbO_2) is a lead IV compound and is relatively unstable. It reacts with sulfuric acid (H_2SO_4) to form lead sulfate. This is a lead II compound and is more stable than lead dioxide.

Lead dioxide
2x PbO_2

Oxygen

Sulfur

Sulfuric acid
2x H_2SO_4

Pb^{2+}

S^{2-}

Hydrogen

Lead sulfate
2x $PbSO_4$

Water
2x H_2O

Oxygen
O_2

The reaction that takes place when lead dioxide reacts with sulfuric acid can be written like this:

$$2PbO_2 + 2H_2SO_4 \rightarrow 2PbSO_4 + 2H_2O + O_2$$

In this reaction, lead dioxide acts as an oxidizing agent—a compound that gives up the oxygen it contains in a reaction.

In this reaction between lead nitrate and potassium iodide, lead iodide is forming and separating out.

these are calcium plumbates (Ca_2PbO_4 and $CaPbO_3$), which contain a lead atom bonded to one or two calcium atoms and several oxygen atoms.

Reactions with oxygen

Lead forms three main compounds with oxygen. The simplest and most important, PbO, is known as litharge and has an orangey color. It has been used since ancient times in pottery glazes and paints.

The bright yellow crystals on the left of this picture are lead chromate ($PbCrO_4$). This is also called chrome yellow and is used as a pigment in paints, including that used for road markings. The orange crystals are potassium dichromate ($K_2Cr_2O_7$).

Reactions with lead dioxide produce oxgyen, so this compound is often used in matches and fireworks, where oxygen is needed for reactions to take place.

Another is red lead or minium (Pb_3O_4), which has similar uses to those of litharge. Lead dioxide (PbO_2) is less stable than the other oxides and gives out oxygen in reactions with acids and other substances.

Reactions with carbon

Many organic (carbon-based) compounds containing lead have also been produced, but the bond between lead and carbon is weak, so most of these are not very stable and break down at temperatures above 212°F (100°C). Lead IV organic compounds are usually more stable than lead II organic compounds because they are generally formed by sharing electrons, rather than gaining or losing them.

Lead chelates

A special characteristic of lead is to form complex compounds, called *chelates*, with several groups of atoms at the same time. Lead chelates are able to dissolve in water, while most simple lead compounds cannot. Forming these complex compounds inside the body is a very effective way of treating people with lead poisoning. The poisoned person is given a substance that forms a chelate with lead, and the chelated lead is then gradually washed out of their system.

Lead poisoning

Lead and its compounds can cause poisoning. They are not particularly strong poisons, so getting a tiny amount of lead inside you would not kill you. Lead can build up over time within the body, however, so even a very low lead concentration in the environment can become dangerous over time.

German composer Ludwig van Beethoven died in 1827. Recent studies on some of his hair suggest that he may have died of lead poisoning.

Lead in the body

Once lead enters the body, it can be taken into the bones and teeth in place of calcium. Although lead is not especially harmful in these places, it acts as a store of the element that can be released slowly to cause damage in the blood and in soft tissues such as the spleen and liver.

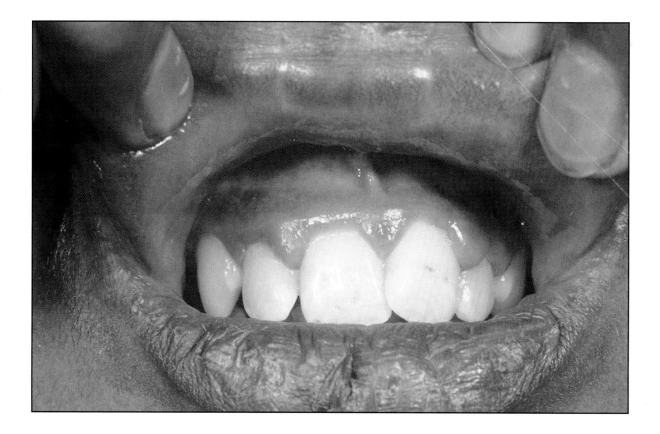

The dark areas on the gums of this patient are a characteristic symptom of severe lead poisoning. They are caused by deposits of lead sulfate ($PbSO_4$).

In the blood and soft tissues, lead prevents enzymes and other essential molecules from functioning properly. In particular, it knocks out some of the enzymes needed to produce hemoglobin (the substance in the blood that carries oxygen). This means not enough hemoglobin is produced, causing a disease called anemia (in which the blood cannot carry enough oxygen to supply the body). At the same time, there is a buildup of the substances made in the process of producing hemoglobin, and these cause some other effects of lead poisoning— headaches, stomach cramps, and constipation. Lead also affects the nerves and brain cells, and once it has reached the brain it is especially difficult to remove. Signs of lead poisoning in the nervous system include irritability, poor hearing, and a lowering of intelligence and physical coordination. If the poisoning becomes even worse, the sufferer becomes confused and paralyzed and may eventually die.

Children are more at risk from lead poisoning than adults. This is because the body of a child absorbs a much higher proportion of any lead that passes through it and, once in the blood, lead causes damage at lower concentrations in children than in adults.

Banned lead

Until relatively recently, people could get lead poisoning from many different sources. In the second half of the 20th century, however, people became more and more aware of how much damage was being done by lead in the environment. As a result, measures have been taken to reduce the

DID YOU KNOW?

LEAD SINKERS

Lead weights used by anglers have killed many wild riverbirds. The lead weights were used as sinkers, to make the bait sink lower in the water, where a fish might be tempted to take it. If any of these sinkers became detached from the fishing line, they would sink to the bottom, where swans and other river birds often search for food. Many birds have eaten lead weights by mistake and died. As a result of this problem, sinkers are no longer made from lead.

amount of lead with which people—especially children—come into contact. The use of household paints containing lead (one of the main hazards) was banned in the U.S. in 1978, and no leaded gasoline has been used since the 1990s. Similar changes have also taken place in European countries.

In addition, doctors are much more aware of the need to test people for lead posioning. The result is that, although the health of some people could still be slightly affected by lead, serious lead poisoning in the U.S. is now rare.

Many wild swans, ducks, and common loons have died from lead poisoning after swallowing just one lead sinker, less than 1 inch (2 cm) in diameter. Today, many anglers use sinkers made from materials such as tin, clay, and steel, which are not poisonous to waterbirds.

Lead in gasoline

L ead is no longer used in gasoline in many countries—in the U.S. it was phased out and finally banned in 1995. This change has occurred only recently, however. Until a few years ago, lead in gasoline was a major source of pollution. So, what were the advantages of adding lead to gasoline that seemed to make this unhealthy practice worthwhile for so long?

Gasoline in automobile engines sometimes burns unevenly. This causes a problem called knocking, in which engine parts rattle against each other and can be damaged. As early as 1916, U.S. engineer Thomas Midgely, Jr., (1889–1944) had experimented with substances that would prevent knocking, and he found one that was very effective—a lead compound called tetraethyl lead.

Since 1975, people have been able to fill up their automobiles with environmentally safer unleaded gasoline at all major gas stations in the U.S.

An environmental problem

Engine knocking was not a big problem with early cars, but things changed when far more powerful engines began to be used. By the 1970s, gasoline manufacturers were adding a few grams of tetraethyl lead in each gallon to ensure that car engines ran smoothly. Tetraethyl lead seemed to be the ideal additive for gasoline, as it also lubricated the engine valves and reduced engine wear even further.

At that time, the quantities of lead that entered the environment through this route were not thought to be harmful. Today, we know better, and tetraethyl lead has been replaced, gradually but completely, by lead-free compounds that are just as effective as antiknock additives.

Uses of lead today

Today lead has lost some of its traditional uses, either for health reasons or because specialized new materials or processes have been developed. Lead is still used for many other things, however. For example, it is used in bullets and solder (an alloy of lead and tin, which is melted and used to join metals together). Lead glass is also used in lenses for cameras and optical instruments, because lead glass lenses bend light much more than those made from other materials.

Lead paints continue to be widely used, although not in the home. In particular,

The Golden Gate Bridge in San Francisco is painted pink. Local people liked the color of the lead-based paint used for its protective first coat so much that the same shade was used for the top coat!

DID YOU KNOW?

RADIATION PROTECTION
One use of lead has arisen only in recent years. Lead is an excellent absorber of X rays and other potentially harmful types of radiation. Lead-lined aprons are used by medical staff and others working with X rays, and lead is often used in the shielding around nuclear reactors. Leaded glass is also used to give protection from radiation, while allowing operators a clear view.

paints containing the lead compounds minium (Pb_3O_4) and calcium plumbate (Ca_2PbO_4) are used—not for their color, but because they are very effective in protecting iron and steel from corrosion.

Battery power

One major application of lead today is in batteries—particularly lead-acid batteries in automobiles. This is where nearly three-quarters of all lead is used, and around a quarter of a billion (250 million) automobile batteries are sold each year.

Unlike the batteries often used to power toys and flashlights, lead-acid batteries are rechargeable. This is why they are used in automobiles—when the vehicle is moving, a simple machine called a dynamo converts some of the movement energy into electrical energy, which is used to recharge the battery. For this reason, car batteries do not usually run down.

Lead crystal is still produced today and contains large amounts of lead—sometimes up to 36 percent.

LEAD-ACID BATTERIES

In any battery, there are two electrodes: one positive and one negative. In a lead-acid battery, the negative electrode is made of lead, and the positive electrode is made from lead dioxide (see below). Between the two electrodes is sulfuric acid, which is made up of sulfate and hydrogen ions (charged atoms or molecules). Each sulfate ion has two negative charges, because it has two extra electrons. Each hydrogen ion has a single positive charge, because it has lost an electron. At the negative electrode, the lead reacts with sulfate ions (SO_4^{2-}) to form lead sulfate ($PbSO_4$). This releases two "spare" electrons, which flow into the electrode. At the positive electrode, the lead oxide reacts with the sulfate and hydrogen ions (H^+) to produce more lead sulfate. To do this, it needs two additional electrons. If the two electrodes are joined by a wire, the lead oxide at the positive electrode can use the two spare electrons from the negative electrode. When this happens, a current of electrons flows between the two electrodes—and through anything connected between them.

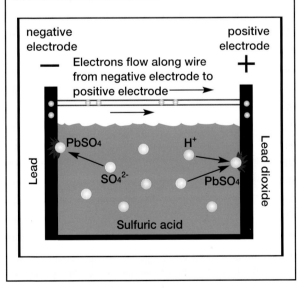

negative electrode **—**   positive electrode **+**

Electrons flow along wire from negative electrode to positive electrode →

Lead

$PbSO_4$ H^+

SO_4^{2-} $PbSO_4$

Lead dioxide

Sulfuric acid

Periodic table

Everything in the Universe is made from combinations of substances called elements. Elements are the building blocks of matter. They are made of tiny atoms, which are much too small to see.

The character of an atom depends on how many even tinier particles called protons there are in its center, or nucleus. An element's atomic number is the same as the number of protons.

Scientists have found more than 110 different elements. About 90 elements occur naturally on Earth. The rest have been made in experiments.

All these elements are set out on a chart called the periodic table. This lists all the elements in order according to their atomic number.

The elements at the left of the table are metals. Those at the right are nonmetals. Between the metals and the nonmetals are the metalloids, which sometimes act like metals and sometimes like nonmetals.

- On the left of the table are the alkali metals. These elements have just one electron in their outer shells.

- Elements get more reactive as you go down each group.

- On the right of the periodic table are the noble gases. These elements have full outer shells.

- The number of electrons orbiting the nucleus increases down each group.

- Elements in the same group have the same number of electrons in their outer shells.

- The transition metals are in the middle of the table, between Groups II and III.

Group I

1			
H Hydrogen 1			
3 **Li** Lithium 7			
11 **Na** Sodium 23			

Group II

4 **Be** Beryllium 9
12 **Mg** Magnesium 24

Transition metals

19 **K** Potassium 39	20 **Ca** Calcium 40	21 **Sc** Scandium 45	22 **Ti** Titanium 48	23 **V** Vanadium 51	24 **Cr** Chromium 52	25 **Mn** Manganese 55	26 **Fe** Iron 56	27 **Co** Cobalt 59
37 **Rb** Rubidium 85	38 **Sr** Strontium 88	39 **Y** Yttrium 89	40 **Zr** Zirconium 91	41 **Nb** Niobium 93	42 **Mo** Molybdenum 96	43 **Tc** Technetium (98)	44 **Ru** Ruthenium 101	45 **Rh** Rhodium 103
55 **Cs** Cesium 133	56 **Ba** Barium 137	71 **Lu** Lutetium 175	72 **Hf** Hafnium 179	73 **Ta** Tantalum 181	74 **W** Tungsten 184	75 **Re** Rhenium 186	76 **Os** Osmium 190	77 **Ir** Iridium 192
87 **Fr** Francium 223	88 **Ra** Radium 226	103 **Lr** Lawrencium (260)	104 **Unq** Unnilquadium (261)	105 **Unp** Unnilpentium (262)	106 **Unh** Unnilhexium (263)	107 **Uns** Unnilseptium (?)	108 **Uno** Unniloctium (?)	109 **Une** Unnillenium (?)

Lanthanide elements

Actinide elements

57 **La** Lanthanum 39	58 **Ce** Cerium 140	59 **Pr** Praseodymium 141	60 **Nd** Neodymium 144	61 **Pm** Promethium (145)
89 **Ac** Actinium 227	90 **Th** Thorium 232	91 **Pa** Protactinium 231	92 **U** Uranium 238	93 **Np** Neptunium (237)

The horizontal rows are called periods. As you go across a period, the atomic number increases by one from each element to the next. The vertical columns are called groups. Elements get heavier as you go down a group. All the elements in a group have the same number of electrons in their outer shells. This means they react in similar ways.

The transition metals fall between Groups II and III. Their electron shells fill up in an unusual way. The lanthanide elements and the actinide elements are set apart from the main table to make it easier to read. All the lanthanide elements and the actinide elements are quite rare.

Lead in the table

Lead has 82 protons in its nucleus, so it has atomic number 82. This element belongs to Group IV of the periodic table, along with carbon (a nonmetal), silicon and germanium (both semimetals), and tin (a metal). Lead is not very reactive and does not corrode easily, because its surface is protected by a layer of lead oxides.

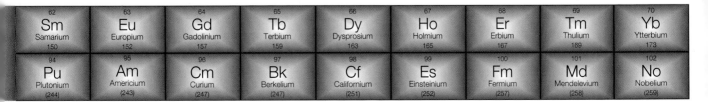

	Metals
	Metalloids (semimetals)
	Nonmetals

| 82 | Pb | Lead | 207 | Atomic (proton) number / Symbol / Name / Atomic mass |

Group VIII

Group III	Group IV	Group V	Group VI	Group VII	2 He Helium 4
5 B Boron 11	6 C Carbon 12	7 N Nitrogen 14	8 O Oxygen 16	9 F Fluorine 19	10 Ne Neon 20
13 Al Aluminum 27	14 Si Silicon 28	15 P Phosphorus 31	16 S Sulfur 32	17 Cl Chlorine 35	18 Ar Argon 40
31 Ga Gallium 70	32 Ge Germanium 73	33 As Arsenic 75	34 Se Selenium 79	35 Br Bromine 80	36 Kr Krypton 84
49 In Indium 115	50 Sn Tin 119	51 Sb Antimony 122	52 Te Tellurium 128	53 I Iodine 127	54 Xe Xenon 131
81 Tl Thallium 204	82 Pb Lead 207	83 Bi Bismuth 209	84 Po Polonium (209)	85 At Astatine (210)	86 Rn Radon (222)

28 Ni Nickel 59	29 Cu Copper 64	30 Zn Zinc 65
46 Pd Palladium 106	47 Ag Silver 108	48 Cd Cadmium 112
78 Pt Platinum 195	79 Au Gold 197	80 Hg Mercury 201

62 Sm Samarium 150	63 Eu Europium 152	64 Gd Gadolinium 157	65 Tb Terbium 159	66 Dy Dysprosium 163	67 Ho Holmium 165	68 Er Erbium 167	69 Tm Thulium 169	70 Yb Ytterbium 173
94 Pu Plutonium (244)	95 Am Americium (243)	96 Cm Curium (247)	97 Bk Berkelium (247)	98 Cf Californium (251)	99 Es Einsteinium (252)	100 Fm Fermium (257)	101 Md Mendelevium (258)	102 No Nobelium (259)

Chemical reactions

Chemical reactions are going on all the time—candles burn, nails rust, food is digested. Some reactions involve just two substances; others many more. But, whenever a reaction takes place, at least one substance is changed.

In a chemical reaction, the atoms stay the same, but they join up in different combinations to form new molecules.

Writing an equation

Chemical reactions can be described by writing down the atoms and molecules before and the atoms and molecules after. Since the atoms stay the same, the number of atoms before will be the same

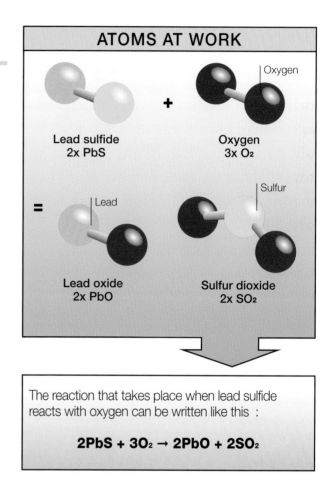

ATOMS AT WORK

Lead sulfide
2x PbS

Oxygen
3x O$_2$

Lead

Lead oxide
2x PbO

Sulfur

Sulfur dioxide
2x SO$_2$

The reaction that takes place when lead sulfide reacts with oxygen can be written like this :

$$2PbS + 3O_2 \rightarrow 2PbO + 2SO_2$$

as the number of atoms after. Chemists write the reaction as an equation. The equation shows what happens in the chemical reaction.

When the numbers of each atom on both sides of the equation are equal, the equation is balanced. If the numbers are not equal, something is wrong. The chemist adjusts the number of atoms involved until the equation does balance.

Once lead oxide has been extracted from galena in the first smelting stage, it is heated in a furnace with charcoal (shown here) to produce impure lead bullion.

Glossary

acid: A substance that can provide hydrogen atoms for chemical reactions.

alloy: A blend of two or more metals.

atom: The smallest part of an element that has all the properties of that element.

atomic mass: The number of protons and neutrons in an atom.

bond: The attraction between two atoms that holds them together.

compound: A substance made of atoms of more than one element. The atoms are held together by chemical bonds.

corrosion: The eating away of a material by reaction with other chemicals, often oxygen and moisture in the air.

density: The mass of a substance in a given volume.

electrode: A material through which an electrical current flows into, or out of, a liquid called an electrolyte.

electron: A tiny particle with a negative charge. Electrons are found inside atoms, where they move around the nucleus in orbits called electron shells.

element: A substance that is made from only one type of atom.

enzyme: A protein molecule that has a specific function within a living cell.

ion: A particle that is similar to an atom but that carries a negative or positive electrical charge.

isotopes: Atoms of an element with the same number of protons and electrons but different numbers of neutrons.

metal: An element on the left-hand side of the periodic table.

molecule: A particle that contains atoms held together by chemical bonds.

neutron: A tiny particle with no electrical charge. It is found in the nucleus of every atom.

nucleus: The center of an atom. It contains protons and neutrons.

ore: A collection of minerals from which metals, in particular, are usually extracted.

organic compound: A compound that contains a chain of carbon atoms.

oxidation: A reaction where oxygen is added to, or one or more electrons are removed from, a substance.

periodic table: A chart of all the chemical elements laid out in order of their atomic number.

proton: A tiny positively charged particle, found inside the atom's nucleus.

radioactivity: The sending out of particles from the center of an atom.

reduction: A reaction where oxygen is removed from, or one or more electrons are added to, a substance.

refining: An industrial process that frees elements, such as metals, from impurities or unwanted material.

valency: The usual number of bonds an atom can form with other atoms.

Index

ML

12/02